THE LIMITATIONS OF DIY WEBSITES

◆ ◆ ◆

LIMITATIONS OF DIY WEBSITES

For Small Business Owners

◆ ◆ ◆

By:

Antonette El Baz

Limitations of DIY Websites

ISBN: 978-0-578-59881-9 (Paperback)

Ordering Information:
Quantity sales. Special discounts are available on quantity purchases by corporations, associations, and others. For details, contact the publisher at the address above.

Printed by Kindle Direct Publishing, in the United States of America.

First printing edition 2019.

www.WebsitesForBusinessOwners.com
antonette@WebsitesForBusinessOwners.com

CONTENTS

PREFACE

♦ ♦ ♦

Should you hire a Web Designer or DIY?

Years ago, I started an accounting/bookkeeping and tax business; I didn't have an abundance of money because I was a new startup with very little capital for my new business. I knew that I needed business cards; they are the bare essentials for getting started, but I didn't have much money to put in them.

I ended up getting 50 business cards. These cards were horrible and gave the wrong first impression of my business. They were flimsy; they didn't have a logo, or color, but it was a card and it got the job done so I thought.

The problem with this thinking is that the message my card was sending is:

"I am serious enough to have a card, but too cheap to get a good one".

To me this describes every Wix and Weebly site I have ever seen. Serious enough to pay for a domain name, too cheap to get a real website.

I'm biased, I am a Web Developer and I dislike these types of tools for use of small businesses. I work with Word-Press (A wonderful CMS) and often help clients by modifying and customizing themes.

The world has changed building each site from the ground up is often excessive. But it needs a good foundation.

An automated software does not have skills for SEO that can be tweaked to your industry. Within safe confines, you get what you pay for.

I'm Antonette El Baz. I'm a Designer who codes. I'm likewise a business owner who has owned a string of different businesses in the professional services, hospitality, and retail industry. I've worked on projects internationally, in-house, and remotely on projects for leading brands, agencies, startups, and charities.

I care about creating world-class, useful, and appealing products that help people and make a distinction. I can be as involved in your project as you need me to be; from the seed of the idea, two sketches, innovative direction, design, copywriting, system design, and even the front-end and WordPress build.

I am now based in Clermont, Florida with my amazing husband, Adil El Baz, 2 children: Amira and Amir, my mom: Pauline Testman who lives with me, and my brother: Anthony Testman, Jr.

After seven years of corporate life as an accountant, eight years of entrepreneurship with several businesses in accounting, retail, and hospitality under my belt, I learned to code and became a remote freelance Web Designer and Developer, and President of Websites for Business Owners.

To understand the pros and cons of the different websites, I wrote this book to provide a framework for learning the limitations of DIY websites. As you weigh your options, I hope that this book will become a primer for small business owners and professionals, helping business owners across the country to learn and practice how to build their brand story, not just a website.

- Antonette El Baz, MBA,
Web Designer
President of Websites for Business Owners

INTRODUCTION

◆ ◆ ◆

I talk to people every day who call or e-mail asking for help. Most of these business owners went with Wix, Weebly, or GoDaddy. They have created a free website or blog, or a low-cost budget service like Vistaprint or whatever.

These businesses are unhappy because they realize that they cannot change the site's design, cannot change text or links or menus, cannot add extra services such as appointment scheduling or eCommerce or without substantial extra costs. Many times their websites will not work on mobile devices (and now more people access the internet via cell phones than do using laptops, so ignoring mobile is a huge oversight).

These cookie-cutter web companies nickel and dime you to pay extra for domain names, to get ads removed, and most times the sites look cheap and tacky and most have no Search Engine visibility.

This means that most will never be found in local Google searches so it wastes your time and money with a product very few people can find or use. It's an online business card showing that the business owner is:

a) cheap

b) out of touch

c) does not care if you would like to learn more about a business through their website

Consumers can and will make other assumptions based on what they see or don't see.

Before we talk about what to look for in a designer, let's talk about the differences between a designer, a developer, and an implementer. We do not set these labels in stone, but knowing them can help you decide what type of professional you need.

A **web designer** is someone who creates the look and mood of your website.

A designer will work with you to choose colors, design branding/logos if needed, discuss layouts for your website's pages, and create mockups (pictures of how the actual website will look).

Web design is more than just making pleasant pictures; a designer should be knowledgeable about concepts like calls to action, organizing the site's content, and setting up layouts that will meet the goals for the site.

A **web developer** is someone who uses a designer's mockups to build a functional website.

If your website is built on WordPress (here's why it should be), the developer may customize an existing theme or template to match the mockups, or may build a custom theme. Developers may add functionality to an existing website or help with things, such as troubleshooting.

An **implementer** may fall somewhere between a designer and developer in terms of what they can do.

This person might work with you to find a theme you admire, install the theme along with any needed plugins, and teach you how to manage your site once it's finished.

Hiring an implementer is often cheaper than working with both a designer and developer, though your pro-ject may need things an implementer cannot offer. Working with an implementer is not automatically a lousy decision–it just depends on what you need and what budget you have available.

CHAPTER 1

Hire or DIY?

◆ ◆ ◆

While there is an abundance of advantages of having a Web designer build your small business website for you, sometimes it can make sense to DIY and build your own website.

If you are a personal blog with a hobby, if you are getting married and looking for a wedding website to share with friends and family, then a website builder may be the best choice for you.

Just keep in mind that you must invest time and hire creative resources (such as photography, videography, graphic design) to create a DIY website which is easy to use.

It makes more sense to hire a Web Designer if your website will rely on revenues, regardless if you're a startup, then investing into a professional website is the way to go.

You want to make sure that your website works well, is easy to navigate, and meets the needs of your online shoppers to maximize revenue opportunities. And a professional can help with that. While your web partner is designing your website, it's a good time to build out your product lines, and plan out your digital marketing initiatives so you can be successful upon launch.

If you're a company who requires a higher level of customization, support, and expertise, then hiring a pro is the best choice. Your website needs to support your brand and drive your business goals.

A professional web designer takes the time to understand your business, your website goals, and has the proven ability to make strategic recommendations on the features and functions of your website to make sure the best experience for your potential customers.

Web design agencies are full of professional designers who have years of experience creating websites for businesses and individuals. That is what they do all day.

They've seen just about every design, style, business, or technical challenge you can throw at them and they know how to create, upgrade, and fix them all. Because of this, they can create a website in a fraction of the time that you could and make it look and operate in a way that most regular folks could never do.

Modern web design is not at all what it was a decade ago. If you were on the internet back then, most pages were static and those for businesses were a simple landing page with maybe a few other pages.

Now it's practically expected that you have headers, footers, a sleek design that moves and changes as the mouse goes over it. The progress happened so that many of us barely noticed. Even using a simple platform such as WordPress you must be able to work with CSS coding to make your website "pop" and be appealing to consumers.

Those who have an abundance of experience will avoid a set of the beginner mistakes which most people make. There is much to be said for experience, with something as touchy as web design. And if you hire a digital marketing agency who also handles web pages, then they've done the analytics on what converts and what does not.

This is often guarded information and a couple of layouts can be the key to an agency being a hit with their customers. Using a professional web design agency gives you the benefit of all the experience that they've already gained.

Creating a website doesn't end when its up, running, and attracting new customers. It's a living thing and needs constant attention to keep it working and growing with your business.

Not only will web designers give you expert advice while designing your website, but they are also there when there are issues and changes that need to be made. Many times website performance will wane due to design or technical updates that need to be made over time.

The web design agency you hire is there for you before, during, and after the website is up and doing it's thing. Agencies are there to offer expert advice on how to improve your website as time goes by and to fix all that goes wrong along the way.

These agencies built your website and recognize what to do if it has issues or needs to be updated to the latest technology.

They are also a convenient, quick call away when you want to change anything on the site and it can be done faster than you could on your own.

No matter what the issue, it's comforting to realize that there are experts on the other end of the phone who can fix it and get your website performing again.

CHAPTER 2

Quality or Low Priced?

◆ ◆ ◆

Hiring a website designer will give you a high quality site. This is one of the biggest reasons to hire a website designer. Sure, there are plenty of free website design templates out there, but they're basic. You can't expect to create a top-of-the-line, unique website with one of these cookie-cutter tools.

A website will require several features, including images, headers, plugins, and codes. All of this may sound like techie jargon to you if you have no information technology/ programming background. But it's all second nature to professional web designers.

A reputable Web designer can create for you a site that is both dynamic and attractive — one that will offer an amazing user experience. Thanks to the technological developments that are continuing to occur in the Web design field, websites today are not like the ones created even three years ago.

Acknowledging that your website is such a major business marketing tool, it only makes sense to hire a designer this year to make sure that yours is attractive and competitive in the modern business landscape.

Another vital part of your website is the content that goes on it. Without quality content your business message will be lost and so will your potential customers.

Using specialized keywords and phrases that are a part of your business and industry are essential to making sure you not only deliver the best products and services to your customers but also so your potential customers see who you are and what you're about.

Quality content makes sure that everything on your website conveys the message and mission to everyone that lands on it. Content not only includes the articles or blogs on your site but also everything from the tag line to the about us page to your contact page.

Every bit of content, large or small, requires in-depth thought and analysis to make sure that your website works to its greatest potential and attracts your business' ideal customers.

Web designers are masters at quality content development and with some input from you about your business, its mission, products/services, and ideal customers, they can customize the content to achieve your business' goals.

Web designers study websites and users' behavior on them. They know what issues or designs will attract users and which ones will repel them.

There are specific things that make a website user-friendly and very important that it incorporates those things into your site. Without them people will get frustrated and leave without finding out what you and your business are about.

With web design, knowing where to put contact forms, addresses, sign up forms, navigation bars, and calls to actions is standard procedure.

The point of a website is to draw people in so they want to delve deeper and find out what you and your business offer them. They want to believe that you can solve their problems.

However, if your website makes it difficult for them to do these things they will find someone else who can make the experience more enjoyable.

These are the things that web design agencies do. Their web designers spend all their time researching and creating websites that attract people and keep them coming back for more.

Web designers recognize what makes the web user's experience pleasant and what turns them off and they use that information to make sure it maximizes your website to attract the most attention from your potential customers.

CHAPTER 3

Custom or Off-the-Shelf?

◆ ◆ ◆

S ite ownership is essential for every single website. Every business should own their own web property and a relationship with their web host company. A company website is just as valuable and important for you to understand and maintain as any other business asset. For example:

Wix.com states:

"Wix does not claim any intellectual property rights over the User Submissions. However, under the Wix.com Terms of Use, you grant Wix worldwide, royalty-free and non-exclusive license to use, distribute, reproduce, modify, adapt, publicly perform and publicly display such User Submissions."

That is scary. By signing up for Wix.com web services, you are allowing Wix.com to change your site and your site's content by agreeing to the Terms of Use.

You should always have control over your website and your content. Period. End of story! Therefore, I feel a way about a product that sets the expectation that anyone can design and develop a website, for free no less.

There are plenty of "out of the box" options available to people today. WordPress is a great example. It has an intuitive interface, you just pick a theme to slap in, fill out the right fields, and within a couple of hours, it will deliver you a functioning website.

Even someone with no experience will not need to spend more than a day or two to get things up and going. It's the custom bits which make a website stand out, however, and that requires coding and extensive tinkering.

Otherwise, you have got a boilerplate website on your hands and that is how you'll be remembered. Going with a professional web design is a great way to make sure you get a website which the competition cannot match.

Most web designers see websites than the businesses they work for. It is an art form and many people overlook that fact, getting caught up all the minute details.

Everyone is creative in their own way, but if you were inclined that way you'd be working as a web designer yourself. The little things can matter a lot in the long run, so why not let someone who has the creative inclination for making great pages take over?

No matter how independent you are. No matter how much spare time you possess, the end result from someone who specializes in professional web design will be leaps ahead of most layman website builders. If you're looking for the best results for your company, then hiring an expert is the way to go.

Web design agencies carry access to professional resources that ordinary folks do not hold. There are tools, web development kits, and add-ons that are only available to professional design agencies.

These resources make it possible for design firms to do things that you could never even dream of doing. Web design agencies acquire the latest up-to-the-moment technology for keeping your website up to date, running at the best speeds, and incorporating the latest designs and bug fixes available.

Resources like these would cost you a fortune to get on your own (if you even could get them) but web design agencies can spread that cost among all of their clients to make it cheaper for you in the long run.

CHAPTER 4

Dependable or Unreliable Design?

◆ ◆ ◆

A nother reason to enlist the help of a professional website designer? A professionally created site is simply more reliable. The risk you run in producing your own site is that some factors may go wrong.

What if you do not understand how to address them? Solving issues with a website can be both time consuming and costly. For instance, you could pay hundreds of dollars for emergency website help and lose valuable time while you're waiting for your issues to be addressed.

All of this can be avoided if you allow a professional designer to create your website and ensure that it is protected from crashing and costly errors.

Running a business is already stressful. Don't add to your stress by forcing yourself to maintain a website you know nothing about.

Allow a professional designer to handle this aspect of your business so you can focus on doing what you well — running your company.

A growing number of Web users are using smartphones and iPads, not just computers, to get online. Therefore, it's critical that your business's website is compatible with today's mobile technologies.

If your website isn't mobile-friendly, your audience will "bounce off" of the website. With the global increase of mobile users and the fact that 80% of internet users own a smart phone, it's no big surprise that a great website it must bear mobile compatibility.

More and more people are accessing the internet via their mobile devices such as smart phones and tablets, and if they do not include a good user experience on your site, they will not come back.

Google states that 61% of users are not likely to return to a mobile site if they had any trouble accessing it, and of those people 40% will visit a competitor's site instead.

That's a lot of eyeballs to lose just because your website wasn't compatible with the latest technology and trends.

Ensuring your site mobile compatible is standard procedure for web design agencies these days. It not only affects how many people a website reaches but it also affects your website's SEO ratings (which I will discuss more on in Chapter 6: Optimize or Plummet?).

A website that is mobile compatible gets higher SEO rankings in all the search engines and that comes back to more eyeballs on your site and your business.

That means they aren't spending much time on your site, and you can't make sales in this situation. The right designer will create your site using responsive design technology so you can make money in the long run.

One of the most annoying that experiences a web user can have is to navigate a slow website. A DIY website that loads will no doubt lose visitors and authority in today's search domains. This will end up costing you many promising business opportunities.

Many sites aren't built to function optimally, but yours

need not be one of them. A professional designer will create a website featuring integrated features and plugins to provide improved security and speed. In this way, your website can easily become the talk of the town for all the right reasons.

By hiring a website designer versus trying to create your own site, you give yourself a valuable competitive advantage in the form of a professional and properly functioning website.

Many small business owners try to put together their own sites to save money.

The problem, though, is that they don't have experience in this area, and it shows. Your professionally created website will far exceed what your competitors hold on the World Wide Web. That will help you remain ahead of your competition in the online race.

A Web designer can deliver to you handy website features. It would take you days to figure out how to add these features to your site on your own. Your designer can help you increase your revenue without your having to lift a finger to make this happen. Trust me, you can't beat that!

CHAPTER 5

Online Strategy or Online Inactivity?

◆ ◆ ◆

A professional developer can create for you a solid strategic plan for your site. Your designer will keep your business model and future organizational goals in mind while creating your site.

You work hard in your business to make sure the services and products you offer are of professional quality, and your website needs to reflect that professionalism too. Like it or not, people decide about you and your business in the first few seconds they visit your website.

A client of ours, a nutrition company was an online store looking to help boost their online sales. These guys are a large business whose website was already experiencing a lot of online traffic and sales.

Websites for Business Owners job was to create a new website that would provide a better user experience for their customers and help boost sales further.

We added some new functions, like a rewards program and a better search function to help customers enjoy the site and use it more. Since launching the new website in June 2019, our client's sales have increased 20%. That's what we call a win!

How your website looks and functions will influence peoples' opinions about you and your business. If your website is sloppy, functions badly, or lacks the information that they are looking for, they will assume that you are lackadaisical in your approach to everything in your business.

Customers will assume, either consciously or unconsciously, that if you didn't care enough to make a decent website that you won't care enough to take care of their needs either.

That decision will send potential customers into the arms of your competitors who do have a professional website whether their products or services are better than yours.

Business owners often read into the future and decide. Do you keep up with web design trends to follow. Take a glance at them and adopted them to stay ahead of your competitors?

People often visit a website and immediately leave the webpage. Can you identify the reason behind this? Probably because they didn't find your website to be attractive enough. And from this, they often assume about the company's web design services.

So, what do you think? How should your website design be? Website design should be such that describes your company and about its products, without having a proper strategy.

How will you influence visitors to stay on your website for longer periods of time and generate more sales. With the changing times, there is a shift in the web design trends. And if you are busy trying to DIY, you might remain aback.

However, a designer's goal will be to establish a solid

foundation for the site so it will thrive long term.

CHAPTER 6

Optimize or Plummet?

◆ ◆ ◆

E veryone has heard about this magical thing called SEO or Search Engine Optimization, but few learn what it is or what it requires to achieve it.

This is another area that web designers excel in. They know what the latest SEO requirements are and how to make sure that your website gets the most eyeballs in the fastest amount of time.

What this means for you is that your site gets found by all the people who need your services or products and that translates into new customers and a better bottom line for your business.

Search engine optimization, or SEO, isn't just a buzz-word these days. It's an important tool for any business owner trying to compete currently and beyond. Your new website needs to be optimized, so it appears in today's top search engines.

Even the best-looking website on the World Wide Web is useless if potential customers cannot find it.

Making your business website search engine friendly means knowing the exact keywords, phrases, and content that will get it recognized by all the major search engines. There are also performance and architecture options that can either kill your visibility or make it shine.

Staying on top of these things requires a seasoned professional. SEO is something that is ever changing and therefore not something that the average business owner has the time to stay up to date on.

A client of ours needed a revamp their online shop selling baby products. Previously, they were receiving one or two online orders a week and their site wasn't ranking for their keywords.

Now, they've told us they receive a couple of orders every day thanks to the revamped site. They also moved up the rankings for the keywords and thrilled about the increased business that the new website has brought to their business.

Another client, a private daycare, had no existing website at all, so we were starting from scratch with this project. This daycare wanted to set up their online presence and get their website to feature when people searched for daycare centers in the Central Florida area.

Websites for Business Owners set up the website for this client and also set up a Google+ page, complete with the Central Florida location of the Cooperative added to Google Maps. The website package included SEO.

We provided meta titles and descriptions for all pages, wrote some new content for the client, and linked the pages internally using keywords. The result? We got OPK to the front page for their most important keywords, beating out competition from older and more established websites! That's a success story and a half!

For professional web designers, SEO is just a regular part of their job and they are outstanding at it.

An expert in design and SEO can increase your website's chances of appearing high in search engine page results. The closer you are to the number-one spot on the first page of results, the more opportunities you gain to draw potential customers in.

CHAPTER 7

Conserve Time or Drain Time?

◆ ◆ ◆

E ven if you are a master web designer in your spare time, most likely you need to be spending your time working in your business not on its website. Even though you might rationalize doing the website yourself as a way of investing in your business, the fact is that the time you spend creating your website will take away from the work you need to be doing in your business.

You can't be in two places at one time. Trading one of these tasks for the other will be a wash as far as time and money. At worst, it will create a loss of time and income that should have been contributing to the bottom line of your company.

Web Designers design websites for a living. They recognize the fastest and most efficient ways to do so and will have your business website up and attracting customers long before most could decide on a website template.

This allows you to work on your business doing things that matter to you. Perhaps one of the most invaluable reasons to hire a professional designer is that this will end up saving you time on a typical business day. And let's face it:

Time is money.

For instance, your designer can provide for you an online quoting system, a form for booking appointments or a contact form that will eliminate the need to book appointments via telephone.

A more efficiently run business leads to more revenue and thus a stronger bottom line.

You could spend all of your time learning how to do this but chances are that you're not a professional Web Designer and the time spent tweaking a page while you're also trying to run a business can be downright exhausting.

Hiring someone who creates sites for a living will save you time. It's pretty much that simple, you hash out what you want and then it's fire-and-forget on your end while you allow the magic to happen.

Most people have their hands full enough with just managing their own side of the business. There's no need to complicate things by having to learn platforms, coding, and then perfecting all of it.

It includes SEO most of the time. Any Web Designer worthy of the name will know at least the basics of technical SEO. Even if you're the one generating the content, it's important that it checks all of those little boxes if you want to get to the top of the SERPs.

Since most organic traffic hits the front page of Google by a wide margin. Your SEO will always be a concern, but a professional Web Designer will ensure that the stuff in the background is functioning while allowing you to focus on producing great content and getting those coveted backlinks.

A website is never an autonomous process. There will

always be issues which need to be sorted, whether it's a security breach or just a malfunction in the code.

A professional Web Designer should be able to help you out if something goes wrong. The less downtime your site goes through, the better off you'll be. It's also easy to turn customers off with things like slow loading speeds.

The day-to-day maintenance of a site doesn't take long. As long as you're not facing serious issues. Unless you're a Web Developer, chances are that serious kinks can take days to work out, costing you valuable time.

CHAPTER 8

Amass Profits or Waste Money?

◆ ◆ ◆

This one goes back to Chapter 7 but besides saving time, hiring a web design agency will save you a ton of money.

Face it, your job as a business owner is to run the business. It's not to create your company's website and a business owner who tries to do that is only taking valuable time and money away from the business they should be running.

Deciding to create your own website will cost your business more money than getting a professional to create the website for you.

Producing a professional website is not as easy as some of these do-it-yourself sites may imply. Those sites can be very limiting in the choices available to you and at worst, after spending a lot of time and energy creating your site it's very possible it won't even function or look anything like you need it to.

You may do the initial research and decide that if you create your own site it will only cost you a few hundred dollars at the most.

However, when you factor in all the time and effort you have to put into it and the issues that can arise, you'll soon calculate that it is costing you way more than you ever wanted to spend.

That number turns out to be more than the cost of hiring a professional web design agency to do a professional website for you.

CONCLUSION

◆ ◆ ◆

These are the reasons that businesses should not rely on website builders like Wix.com for their website. Design agencies gain access to professional resources that ordinary folks do not hold.

There are tools, web development kits, and add-ons that are only available to professional design agencies and these resources make it possible for design firms to do things that you could never even dream of doing.

Web design agencies own the latest up-to-the-moment technology for keeping your website up to date, running at the best speeds, and incorporating the latest designs and bug fixes available.

These resources would cost you a fortune to get on your own (if you even could get them) but web design agencies can spread that cost among all of their clients to make it cheaper for you in the long run.

These are the limitations of a DIY business website and reasons to hire a web design agency to create your business website and there are many more too. When you became a business owner, it was intending to create a quality business and earn a good income.

Part of that is knowing what things to do yourself and what things you should delegate to others. Building a website, unless your business is website creation, is something you should leave to professional web design agencies so you can spend your valuable time running your business and achieving your goals.

You may ask, but what if I already have a DIY website on a platform like WIX? Don't worry! You are not out of luck. Unfortunately, you cannot just move your site as is to a new hosting company because you don't own your site, but you can have your website professionally designed and implemented on a proven platform like WordPress.

When you are ready to hire a professional web design agency, contact **Websites for Business Owners** by going to our website at:

www.WebsitesForBusinessOwners.com.

Let us help you get your business where you want it to be.

◆ ◆ ◆

APPENDIX

DIY vs Web Designer

◆ ◆ ◆

	DIY	Web Professional
Development experience	For Website Builders: no coding experience required For WordPress: some coding knowledge is required	You don't need experience with web building platforms
Time	You will be required to invest a lot of time You need to prioritize your time for planning, building and QAing the site as well as content development (as well as your regular day-to-day tasks)	You can invest less time by hiring a developer Spend more time on developing content and other marketing channels, not to mention growing your business
Budget	Starting at $50/year for domain & hosting May need additional budget for content, logo, one-time development assistance etc.	Starting at $50/year for domain & hosting For a good developer to build your website: several hundreds to thousands of dollars. Ongoing development work: hundreds to thousands of dollars
Customization	Customizations will be limited by your skills and potentially the website builder you use	Customization is limitless (when developer builds site on WordPress)
Content writing, logo design & additional website-related tasks	Complete yourself at minimal/no cost Hire a freelancer through a marketplace at relatively low cost	Professional content writer will charge an average of $50 per 500 words Professional logo design will start in the hundreds of dollars